simple machines

# Pulleys and Gears

RIGBY
INTERACTIVE
LIBRARY

David Glover

This edition © 1997 Rigby Education
Published by Rigby Interactive Library,
an imprint of Rigby Education,
division of Reed Elsevier, Inc.
500 Coventry Lane
Crystal Lake, IL 60014

Printed in Hong Kong / China

01 00 99
10 9 8 7 6 5 4 3 2

**Library of Congress Cataloging-in-Publication Data**
Glover, David.
    Pulleys and gears/David Glover.
        p.    cm. — (Simple machines)
    Includes index.
    Summary: Introduces the principles of pulleys and gears as simple machines, using examples from everyday life.
    ISBN 1–57572-084–1 (lib. bdg.)
    1. Simple machines—Juvenile literature.  2. Pulleys—Juvenile literature.  3. Gearing—Juvenile literature. [1. Pulleys. 2. Gearing.] I. Title. II. Series.
TJ147.G57 1997
621.8 ' 11—dc20

                            96–15815
                                CIP
                                AC

Designed by Celia Floyd and Sharon Rudd
Illustrated by Barry Atkinson (pp. 8, 9, 11, 17) and Tony Kenyon (p. 7)

**Acknowledgments**
The publisher would like to thank the following for permission to reproduce photographs:
Trevor Clifford, pp. 4, 5, 12, 14–19, 21, 23; Collections/Keith Pritchard, p. 9; Mary Evans Picture Library, p. 10; Stockfile/Steven Behr, p. 22; TRIP/H. Rogers, p. 13; Skip Novak PPL, p. 7; Zefa/Damm, p. 6; Zefa/Kurt Goebel, p. 20; Zefa/G. Mabbs, p. 21.

Cover photograph by Trevor Clifford

Every effort has been made to contact copyright holders of any material reproduced in this book. Any omissions will be rectified in subsequent printings if notice is given to the publisher.

**Note to the Reader**
Some words in this book are printed in **bold** type. This indicates that the word is listed in the glossary on page 24. This glossary gives a brief explanation of words that may be new to you and tells you the page on which each word first appears.

# Contents

# What Are Pulleys and Gears?

Pulleys and gears are two kinds of special wheels that help make things move.

When you turn the handle on this toy windmill, the sails turn, too. That is because the handle and the sails are connected by a rubber band called the **drive belt.** The drive belt is stretched over two pulleys. Turning the handle moves the rubber band. This makes both pulleys turn.

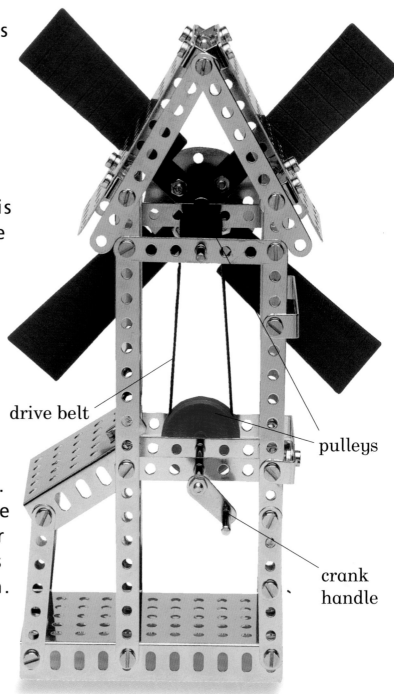

drive belt

pulleys

crank handle

The crank handle of this model is linked to the wheels by two gear wheels. The gear wheels have teeth around their edges. Some of the teeth on one wheel fit between some of the teeth on the other wheel. This is called **meshing**.

When one gear wheel turns, its teeth push the teeth on the other gear wheel. This makes the second gear wheel turn as well.

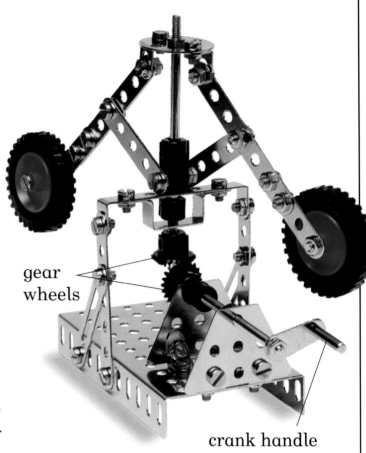

gear wheels

crank handle

## FACT FILE

### Different directions

When two pulleys are connected by a drive belt, they go around in the same direction. Two gear wheels with meshed teeth go around in opposite directions.

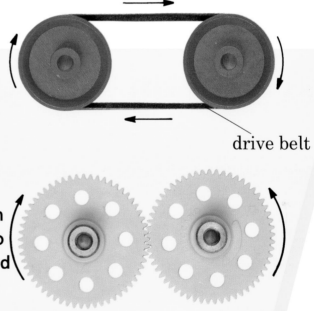

drive belt

# Up the Flagpole

To make a flag go up a flagpole, you pull down on the rope. How can pulling down make something go up?

The answer is at the top of the pole. The rope that you pull goes up over a pulley. It hangs down the other side, where it is connected to the flag. So, as you pull down on one side of the rope, the flag on the other side goes up.

Can you spot the pulleys on this boat? Pulleys help sailors raise the sails up the mast. The sailors pull on ropes on the deck to make the sails go up.

### The first pulleys

No one knows who invented the pulley, but it was probably just a smooth tree branch. It might have been as simple as a rope thrown over a tree branch to lift a heavy load or to put it on a cart.

# Cranes and Block and Tackle

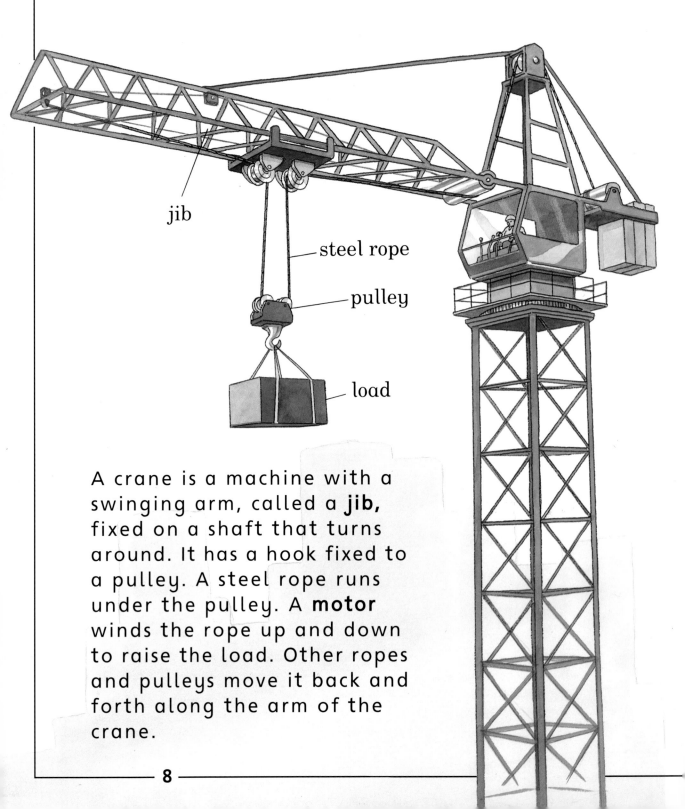

jib

steel rope

pulley

load

A crane is a machine with a swinging arm, called a **jib,** fixed on a shaft that turns around. It has a hook fixed to a pulley. A steel rope runs under the pulley. A **motor** winds the rope up and down to raise the load. Other ropes and pulleys move it back and forth along the arm of the crane.

A block and tackle is a set of pulleys that work together. One person can lift a heavy weight with a block and tackle. This boat was lifted out of the water using a block and tackle.

## Making it easy

With two pulleys you can lift twice as much weight as you could with one pulley and the same amount of pull.

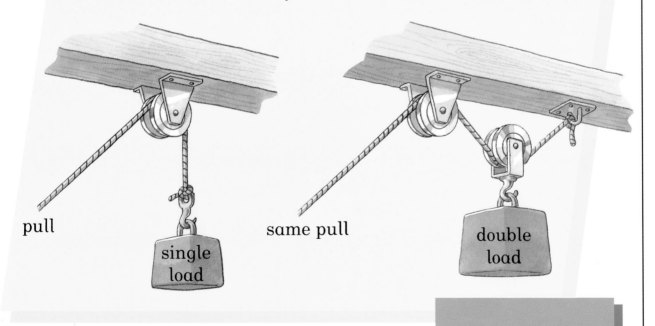

pull

single load

same pull

double load

# Drive Belts

A steam engine drives this old fashioned sewing machine. The engine is joined to the sewing machine by pulleys and a drive belt. The pulley on the machine is smaller than the pulley on the engine. This makes the sewing machine turn faster than the engine.

These pulley wheels are linked by a drive belt. Two wheels of the same size turn at the same speed.

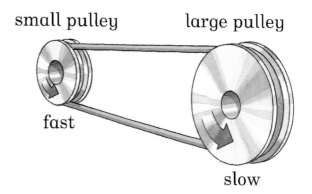

same size – same speed

When the wheels are different sizes, the smaller wheel turns faster than the bigger one.

small pulley

large pulley

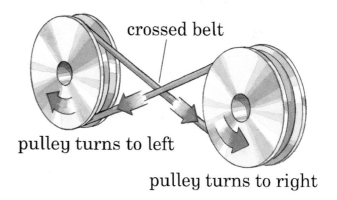

fast

slow

You can make pulley wheels turn in opposite directions by twisting and crossing over the drive belt.

crossed belt

pulley turns to left

pulley turns to right

# Power Pulleys

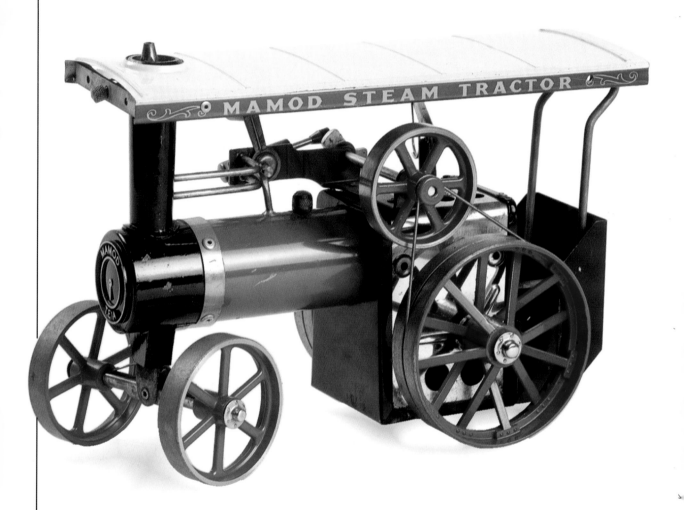

Most of the rides at a carnival go round
and round. Long ago, some of them were
worked by **steam engines** with pulleys
and drive belts. This is a model of an old
steam engine. It has a drive belt that
turns an **electric generator.** The belt runs
between a pulley on the engine and a
pulley on the generator.

This machine crushes sugar cane to squeeze out the juice. The machine is powered by a steam engine and has many moving parts. Drive belts and pulleys make the parts go around.

## FACT FILE

### Be safe!

Drive belts and pulleys are dangerous. You can trap your hair, your hands, or your clothes between the pulley and the belt. All modern pulleys are made safer with special safety guards.

# Gear Kits

You can learn about gears by making models with a gear kit. Flat, round gears are called **spur gears**. One spur gear can turn several other gears. This is called a *gear train.* The gears in the train go around in different directions. If the gears are different sizes, they go around at different speeds.

A gear wheel that looks like a screw is called a **worm gear**. When a worm gear turns, it makes a large spur gear go around very slowly.

Two gear wheels can be connected with a chain that fits over their teeth, like the gears on a bicycle. The chain makes both gear wheels go around in the same direction.

## FACT FILE

### Count the teeth

These three gear wheels have 7 teeth, 10 teeth and 14 teeth. Which two gears would you choose to make one gear turn twice as fast as the other? Look at page 24 to see if you are right.

# Drills, Whisks, and Reels

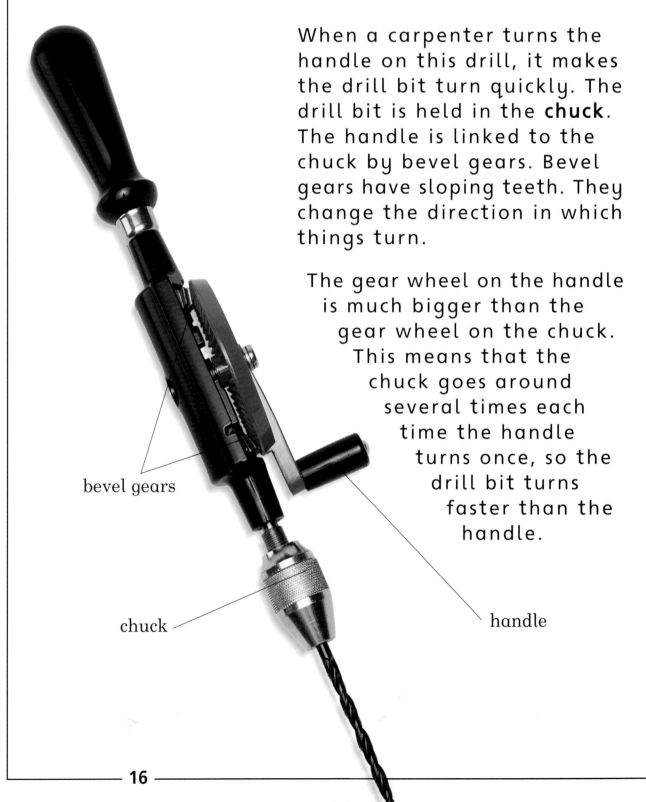

When a carpenter turns the handle on this drill, it makes the drill bit turn quickly. The drill bit is held in the **chuck**. The handle is linked to the chuck by bevel gears. Bevel gears have sloping teeth. They change the direction in which things turn.

The gear wheel on the handle is much bigger than the gear wheel on the chuck. This means that the chuck goes around several times each time the handle turns once, so the drill bit turns faster than the handle.

bevel gears

chuck

handle

Three gears turn the two blades on this whisk. Two gear wheels lie on either side of the big gear wheel on the handle. The blades turn in opposite directions, so food is mixed very well.

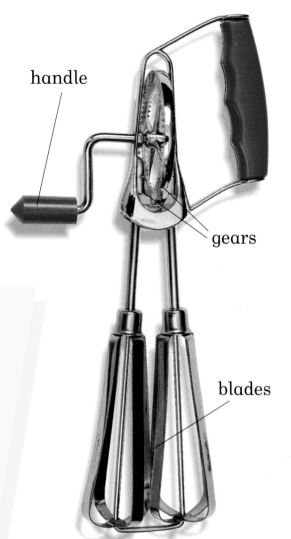

handle

gears

blades

## FACT FILE

### Changing direction

Gears inside this fishing reel work together to change the direction of the handle's turn. As the handle turns, the reel winds in the line.

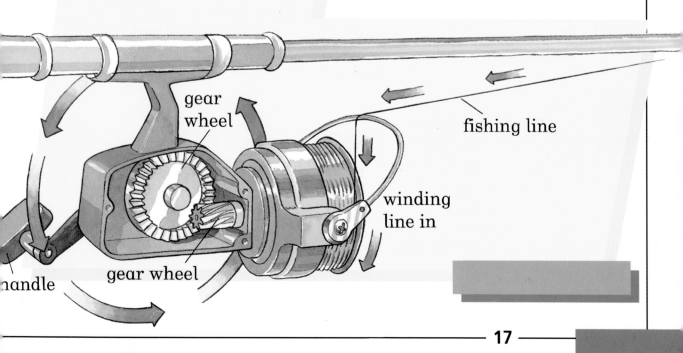

gear wheel

fishing line

winding line in

gear wheel

handle

# Clocks and Watches

hour hand

minute hand

second hand

The back of this watch is open to show the gears.

The three hands on a clock or watch go around at different speeds. One **mechanism** turns them all. Each hand of the clock is linked to the mechanism by different gears.

During the time the hour hand turns one complete circle, the minute hand turns 12 times. Extra gears make the hour hand go more slowly than the minute hand.

Gears also turn the hands of a cuckoo clock. They are powered by falling weights. Extra gears turn the parts that make the cuckoo pop out every hour.

# Mills

gears

grindstone

A windmill is powered by sails and huge gear wheels. The gear wheels turn around and carry the turning force of the sails to the grindstones. Grindstones are stone wheels that grind wheat and other grains into flour.

The gear wheels in this water mill are made from metal. Metal gears are very strong, but they must be oiled to make them run smoothly.

The gear wheels in this toy mill are made from plastic. Plastic gears turn smoothly without oil, but they are not as strong as metal ones.

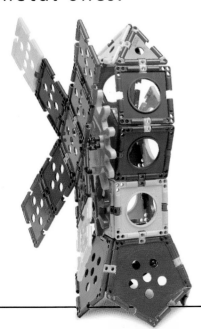

## FACT FILE

### Machine power

Because they used gears, windmills and water mills were among the first machines that weren't powered by people or animals. They were invented more than 1,000 years ago.

# Mountain Bikes

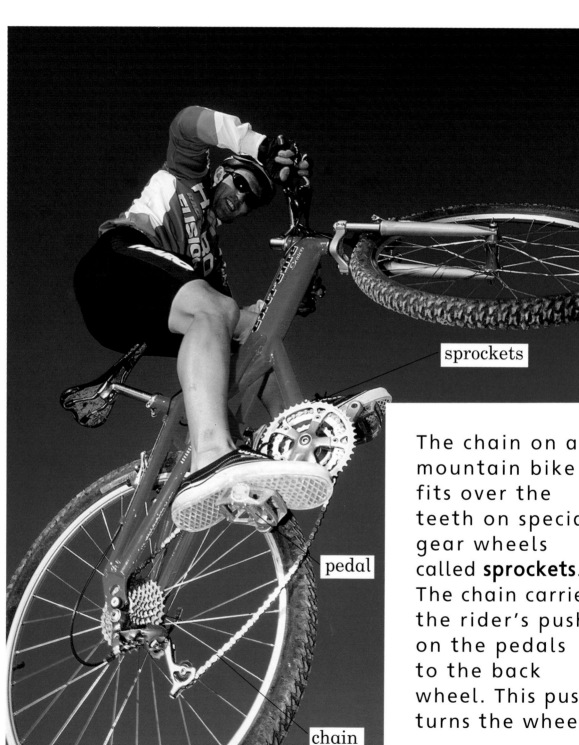

sprockets

pedal

chain

The chain on a mountain bike fits over the teeth on special gear wheels called **sprockets**. The chain carries the rider's push on the pedals to the back wheel. This push turns the wheel.

When a bike's gears are changed, the chain moves between different sized sprockets on the back wheel. Riders shift it to a large sprocket to easily climb hills. They shift to a smaller sprocket to go fast on flat ground.

## FACT FILE

### How many gears?

Some mountain bikes have 28 gears! Track racing bikes have only one gear.

# Glossary

**chuck**   Part on a drill that grips the different sized drill bits   **16**

**drive belt**   Loop of leather or rubber that connects one pulley wheel to another   **4**

**electric generator**   Machine that changes energy from movement into electricity   **12**

**jib**   Long arm on a crane   **8**

**mechanism**   Machine   **18**

**meshing**   When the teeth on two gear wheels fit together   **5**

**motor**   Machine that uses electricity or fuel to make things move   **8**

**sprockets**   Toothed wheels on the pedals and back wheel of a bicycle   **22**

**spur gears**   Flat, circular gears with teeth around the edge   **14**

**steam engines**   Motors or engines that use steam from boiling water to make things move   **12**

**worm gear**   Gear like a screw with a spiral thread running around its surface   **14**

# Index

# Further Readings

Barton, Byron. *Machines at Work*. New York: HarperCollins, 1987.

Taylor, Barbara. *Get It In Gear! The Science of Movement*. New York: Random, 1991.

**Answer to page 15:** the 7 and the 14 because the 7-toothed wheel goes twice as fast as the 14-toothed wheel.